Milk and Milk Products

Instant Notes

Aparna Bhatt

Preface

In India, dairying is being practiced since time immemorial. Since independence Indian dairy industry has made swift progress. In the mean while, dairy science has been emerged as a new subject dealing with the processing of milk and manufacturing of milk products on an industrial scale. This small book is a sincere attempt to provide some basic insights into current milk production and processing practices. Techniques mentioned in this book are strictly in an Indian context. Production techniques and all other information have been written keeping in mind the present scenario of production and processing practices in an Indian dairy industry.

Table of Contents

1. Introduction

MILK is the normal mammary secretion derived from complete milking of healthy milch animal. It is also defined as the whole, clean, fresh, lacteal secretion obtained by complete milking of one or more healthy milch animals. Milk has been evolved as an ideal source of nourishment to the young ones of a particular species producing it. In many parts of the world, cow is the major source of milk for human consumption. However, some other sources of milk like buffalo, goat, sheep, camel etc have also been employed for milk consumption. In India, the principal sources of milk are cow and buffalo. Moreover, in few parts of the country some amount of goat milk is also being consumed.

India ranks high in milk production with a largest livestock population in the world. The average production and per capita availability of milk in India is given below:

Milk production in India		
Year	Production (Million Tons)	Per Capita Availability (gm/day)
2008-09	112.2	266
2009-10	116.4	273
2010-11	121.8	281
2011-12	127.9	290

Source: Department of Animal Husbandry, Dairy & Fisheries, Ministry of Agriculture, GoI

The term 'market milk' refers to the fluid whole milk that is sold to individuals usually for direct consumption. In India, generally the term 'milk' refers to cow or buffalo milk, or a combination of the two. The proportion of cow, buffalo and goat milk to the total milk production in India is given in the table below:

Type of milk	Total production in India (%)
Cow	33.6
Buffalo	63.6
Goat	2.8

Source: F.A.O. Production Year Book, 1974.

2. Nutritive Value of Milk

The story of milk goes back to the beginning of civilization itself. Even in prehistoric times cattle were domesticated for milk, ploughing and other such purposes but at the same time were considered sacred and even worshipped in India. Milk is an ideal food having a very high nutritive value and is also one of the most essential of all foods. It is one of the most complete single foods available in nature for health and promotion of growth. Milk is often regarded as being nature's most complete food. It is one of the most crucial parts of our diet as it provides many of the nutrients which are essential for the growth of the human body. Being an excellent source of body building proteins, vitamins and minerals, particularly calcium milk has gained special importance in our diet at different stages of life.

Fermented-milk products such as yoghurt and soured milk contain bacteria from the *Lactobacilli* group. This bacterium is present naturally in the digestive tract and has a cleansing and healing effect. Hence, the introduction of fermented products to our diet can help prevent certain yeasts and bacteria which may cause illness. Many people suffer from a condition known as 'lactose intolerance' which means that they are unable to digest the milk fat (lactose). Such people can, however, tolerate milk if it is fermented to produce foods such as yoghurt. During fermentation, lactose of milk gets broken down through lactic acid producing bacteria, and by doing so eliminate the cause of irritation.

Proteins: Milk proteins are complete proteins of high quality and contain all essential amino acids in fairly large quantities. Protein is vital for living organisms and being a good source of proteins milk is essential in the diet.

Minerals: Milk is an excellent source of calcium and phosphorus. Both of these minerals together with vitamin D are essential for bones. Milk is however low in iron, copper and iodine.

Vitamins: Vitamins are necessary for normal growth, health and reproduction of living organisms. Milk is a fairly good source of vitamin A, vitamin D (provided the cow is exposed to enough sunlight), thiamine, riboflavin etc. However, it is deficient in vitamin C.

Fat: Milk fat is not only a good source of energy but also contains significant amounts of essential fatty acids accessory for the body. It plays a significant role in the nutritive value, flavor, richness and physical properties of milk and milk products. It is rich source of calories and imparts smooth texture and rich taste to dairy products.

Lactose: Lactose is also known by the name milk sugar. The principal function of lactose is to supply energy. It also facilitates assimilation in the intestine and also checks the growth of proteolytic bacteria by establishing a mildly acidic reaction in the stomach.

Energy value: The energy providing milk constituents and their compositions are as follows:

Milk fat – 9.3 C/g

Milk protein – 4.1 C/g

Milk sugar – 4.1 C/g

*On an average, cow milk furnishes 75 C/100g and buffalo milk 100 C/100g.

3. Composition of Milk

The major constituents of milk are water (85.5-88.5%), fat (3-6%), protein (3-4%), Lactose (~5%), ash or mineral matter (~0.7%). All the solids present in milk are known as 'total solids' (11.4-14.5%) and total solids without fat are referred to as 'milk-solids-not-fat' or 'solids-not-fat.' The minor constituents of milk are phospholipids, vitamins, enzymes, pigments etc.

The range of quantitative differences in milk composition enormously varies from species to species. Such differences have also been recorded in milk from the same source depending upon various factors affecting the composition of milk like, breed of animal, age, stage of lactation, feeding, condition of animal at calving, time of milking, frequency of milking, administration of drugs and hormones and so on. Composition of milk from different animal species is given below:

Composition of milk from different species (per cent)					
	Water	Fat	Protein	Lactose	Ash
Buffalo	84.2	6.6	3.9	5.2	0.8
Cow	86.6	4.6	3.4	4.9	0.7
Goat	86.5	4.5	3.5	4.7	0.8
Human	87.7	3.6	1.8	6.8	0.1

Source: *Outlines of Dairy Technology* Sukumar De (2013).

3.1. Major components:

a) **Water:** It is the medium for suspension of all other components. It constitutes the major proportion of milk most of which is present in free form.

b) **Milk fat:** Milk fat is highly nutritive and is of great economical value. The flavor of milk to a large extent depends upon milk fat. Milk consists of minute globules with an average size of 2-5 microns. It is an oil-in-water type emulsion and is surrounded with an adsorbed layer of fat globule membrane. This membrane carries phospholipids and proteins which forms a complex and stabilizes the emulsion. However, this emulsion gets broken while undergoing certain processing operations like agitation, freezing or heating. On the contrary when milk is held undisturbed, fat globules rise up to the surface and forms a creamy layer. Buffalo milk has a thickest cream layer as compared to the other milks as it contains higher fat content and relatively larger fat globules. However, the size of fat globules depends also on the breed of animal and their lactation period.

c) **Milk protein:** Mammary glands are responsible for the synthesis of milk proteins. Milk protein is a heterogeneous mixture. Along with other milk components it also contains about 0.5 % nitrogen. Out of the total nitrogen present in milk 95% is the milk protein and rest is the non-protein nitrogen. Proteins carry a large number of essential and non-essential amino acids. The major portion of milk protein consists of casein, β-lactglobulin and α-lactalbumin. Milk is an only source of casein which exists in the form of calcium caseinate phosphate complex. Casein is a good repository of several essential amino acids with a highest concentration of glutamic acid. Apart from casein, β-lactglobulin and α-lactalbumin together are known as whey or serum protein. Rennin is an extremely powerful clotting enzyme secreted in the fourth stomach or abomasum of young calves. Its acts as a proteolytic enzyme and causes rapid clotting of milk.

d) Milk sugar or Lactose: In milk, sugar is present in the form of Lactose (a reducing sugar) which on hydrolysis gives glucose and galactose. Milk is the only source of lactose in diet. It is one-sixth of the sweetness of sucrose. When crystallized, it occurs in two forms α-lactose (below 93.5 °C) and β-lactose (above 93.5 °C). α-Lactose imparts a gritty feel and is sparingly soluble when placed in mouth. On the other hand, β-lactose is sweeter and more soluble when placed in mouth. This property of milk is highly important in the preparation of milk based confectionary and other dairy products. Moreover, lactose is fermented by bacteria to form lactic acid which is further used for producing cultured milk products like yoghurt, butter milk etc.

e) Ash or mineral matter: The mineral matter of milk is obtained after ashing of a known quantity of milk at 500-550 °C for a certain period of time until a white residue is obtained which is known as ash or mineral matter. It consists of appreciable amounts of salts of sodium, potassium, calcium, phosphorus, magnesium, sulphate, chloride etc.

3.2. Minor components:

a) Phospholipids: Being an important component of the fat globule membrane, phospholipids contribute to the richness and flavor of milk. It is highly sensitive to oxidative changes and acts as an excellent emulsifier. Its emulsifying property attributes to stabilize the milk fat emulsion.

b) Pigments: The pigment carotene is responsible for the yellow color of milk, cream, butter, ghee etc. Carotene is fat soluble pigment and is reddish- brown in appearance. It is also an antioxidant and acts as a precursor of vitamin A. It is found in higher amount in cow milk ($30/\mu g/g$) as compared to buffalo milk (0.25-$0.48/\mu g/g$) attributing yellowish appearance to cow milk.

c) Enzymes: Enzymes are biological catalysts which are responsible for enhancing or retarding a chemical reaction. The important enzymes found in milk are:

i. Lipase- It is responsible for splitting fat in milk thus results in production of rancid flavor. It can be inactivated by pasteurization.

ii. Phosphate- It is capable of splitting certain phosphoric acid esters which is a basis of phosphatase test for checking pasteurization efficiency.

iii. Peroxidase and Catalase: It is subjected to the decomposition of hydrogen peroxide.

d) Vitamins: Vitamins are vital for the normal growth and functioning of living organisms. Milk is quite a good source of vitamin A.

4. Properties of Milk

i. Acidity- Fresh milk when comes in contact of litmus paper turns blue litmus to red and vice versa. The titrable acidity of cow milk varies from 0.13-0.14 per cent while that of buffalo milk varies from 0.14-0.15 per cent.

ii. pH- The pH of freshly drawn cow milk varies from 6.4-6.6 while that of buffalo milk varies from 6.7-6.8.

iii. Specific gravity- Milk is heavier than water. The specific gravity of cow milk ranges from 1.028-1.030 while that of buffalo milk ranges from 1.030-1.032.

iv. Freezing point- Milk freezes at a temperature slightly lower than that of water. Freezing point of cow milk is -0.547 while that of buffalo milk is -0.549. If the freezing point value is observed lower than the normal, it indicates that the milk is adulterated with water.

v. Color of milk- Cow milk is observed to have yellowish white color while buffalo milk is creamy white in color. Carotene is mainly responsible for that yellowish tinge in milk. Moreover, skim milk is bluish and whey is greenish yellow in color.

5. Milk Processing

Milk processing involves a series of vital steps carried out after milking to ensure fresh and safe milk to the customer. It helps in producing milk with rich flavor and good keeping quality. Milk processing operations involve clarification, pasteurization and homogenization.

i. Receiving milk: The reception of milk in a milk plant or dairy is usually done with the help of large cans or tankers. It is very important to receive the large volumes of milk in a clean and hygienic place as soon as possible because delays may permit deterioration of milk. It should be received continuously within the scheduled period so that various operations in the plant may not be interrupted. Before entering into the plant, milk is subjected to a series of certain mandatory tests called platform tests.

The 'Platform Tests' includes all those tests which are performed to check the quality of incoming milk on the receiving platform so as to make a quick decision regarding acceptance or rejection of milk. The various platform tests are discussed below:

Smell/Odor- This is an excellent indication of the organoleptic quality of milk and can be ascertained very quickly. Usually a trained milk grader is employed for the purpose that ensures whether the milk is fresh and free from off-odors or not. As per his results the milk is accepted or rejected in the platform.

Appearance- After odor test, milk cans are observed randomly to ensure that milk is free from unwanted floating matter, off-color and churned fat globules.

Temperature: The temperature at which the milk is received in the platform should be 5C or below. It should be sufficiently cold to ensure freshness.

Sediment: It shows the visible extraneous material present in the milk. It need not be done daily but should be performed often to ensure clean and safe milk supply.

Acidity: Milk is slightly acidic in nature. The acidity of cow milk is 6.6 while that of buffalo milk is 6.7. The natural acidity of milk does not tend to jeopardize its quality but the acidity developed later may tend to do so.

Lactometer reading: This test is performed to check the adulteration in milk. Addition of water to milk decreases its lactometer reading.

ii. Pre heating: After the acceptance of milk, it is subjected to heat before any further operation so that the flow of milk may not be hampered. This is because as the temperature of milk is raised its viscosity decreases.

iii. Filtration/Clarification: This is done for removing the visible foreign material present in milk which may find their way during its handling. It helps in improving the aesthetic value of milk which may later lead to consumer complaints. In filtration, the suspended foreign particles of milk are removed through straining while clarification is based on the principle of centrifugal sedimentation for removing the suspended dirt particles. To get rid of such impurities milk is generally passed through a centrifugal clarifier which helps in removing dirt and filth. The clarified or filtered milk is now ready for pasteurization.

iv. Cooling and Storage of Raw Milk: Cooling of milk is necessary to avoid deterioration which may occur because of bacterial contamination. Generally cooling and storage of milk is done at 5°C. Various cooling equipments are being employed for proper chilling of milk like; surface cooler, plate cooler, internal tubular cooler, jacketed vats etc. After getting cooled properly milk is than stored in large tanks for its further processing.

v. Standardization of Milk: Standardization of milk refers to the adjustment, i.e., raising or lowering of fat and/or solids-not-fat percentages of milk to a desired value so as to confirm to the prescribed legal requirements. Milk is standardized by the addition of milk or cream with a higher or lower fat percentage as per the need. This can be done by using Pearson's square scheme which is used to calculate the relative quantities of materials involved in the standardization procedure.

vi. Pasteurization of Milk: The term pasteurization has been coined after the name of the scientist Louis Pasteur from France. The term pasteurization refers to the heating of milk to at least 63°C for 30 minutes or 72°C for 15 seconds. After pasteurization the milk is immediately cooled to 5°C or below. The objective of pasteurization is to render milk safe for human consumption and to improve the keeping quality of milk.

Pasteurization is the temperature-time effect which is subjected to kill the most heat resistant bacteria present in milk i.e., *Coxiella burnetii* so that the objective of killing all pathogens can be achieved. In addition to this, pasteurization also aims at inactivating phosphatase enzyme present in milk. This means that adequately pasteurized milk should give a negative phosphatase test so as to ensure that milk is safe for consumption.

vii. Homogenization of Milk: Homogenization is the process of forcing the milk through a small aperture of a homogenizer at 2500-3000 psi pressure with the objective of sub-dividing the fat globules to a size of 2 microns or less in diameter. It gives a brighter appearance and rich flavor to the milk. The milk undergone homogenization does not churn by rough handling and produces soft curd.

viii. Bottling/Packaging: The pasteurized and cooled milk is promptly bottled or packaged to protect the milk against contamination and helping the sale and distribution of milk.

6. Classes of Milk in India

The market price and flavor of milk depends upon its fat content. The variation in milk composition obtained from different sources is an important consideration during milk-processing operations. According to the Prevention of Food Adulteration (PFA) Act, different classes of milk as per their milk fat and milk-solids-not-fat content in India are given below:

PFA standards for different classes of milk in India (%)		
Class of milk	Milk fat	Milk-solids-not-fat (MSNF)
Cow milk	4.0	8.5
Mixed milk	4.5	8.5
Standardized milk	4.5	8.5
Recombined milk	3.0	8.5
Toned milk	3.0	8.5
Double toned milk	1.5	9.0
Skimmed milk	Not more than 0.5	8.7
Full cream milk	6.0	9.0

Source: *Foods Facts and Principles* by Manay and Shadaksharaswamy (2008).

7. Special Milks

a) Sterilized milk: Milk which has been heated to a temperature of 100°C or above for such lengths of time that it remains fir for human consumption for at least 7 days at room temperature is called sterilized milk.

Flow diagram for the manufacture of sterilized milk:

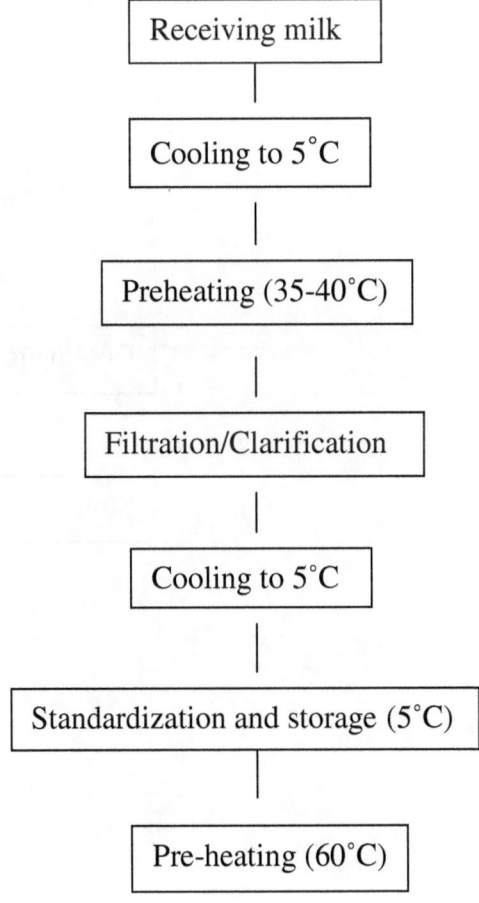

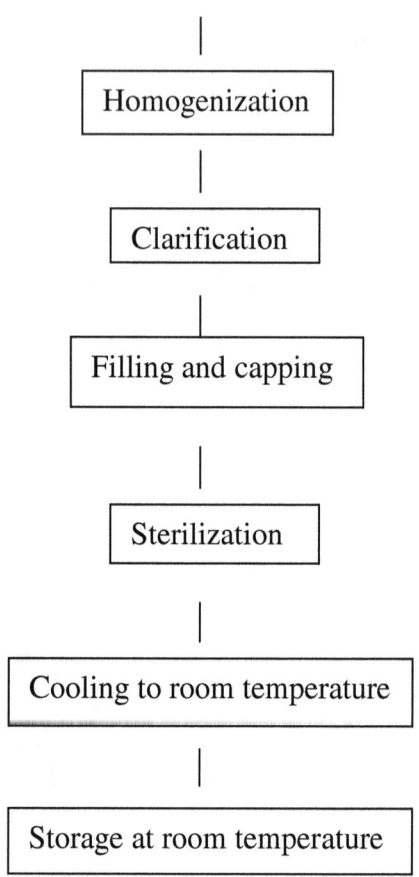

Homogenization

Clarification

Filling and capping

Sterilization

Cooling to room temperature

Storage at room temperature

16

b) Homogenized milk: Homogenization is the process of forcing the milk through a homogenizer with the objective of sub-dividing the fat globules. Homogenized milk ensures breakup of the fat globules to such an extent that no visible fat separation will be seen even after about 48 hours.

Flow diagram for the manufacture of homogenized milk:

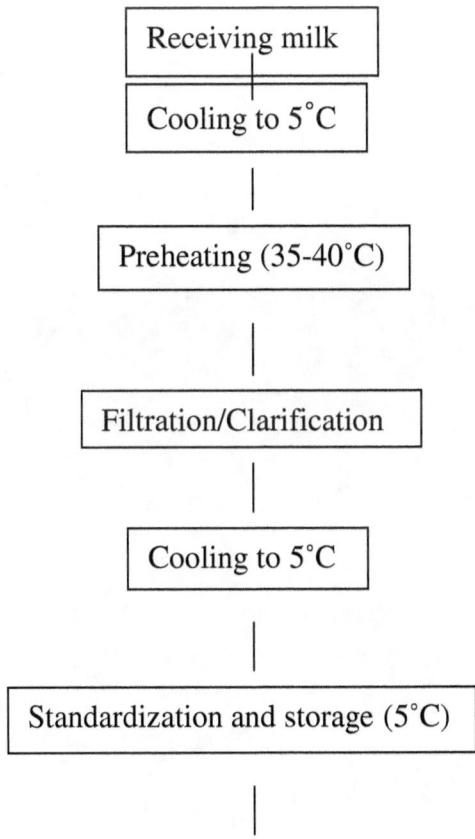

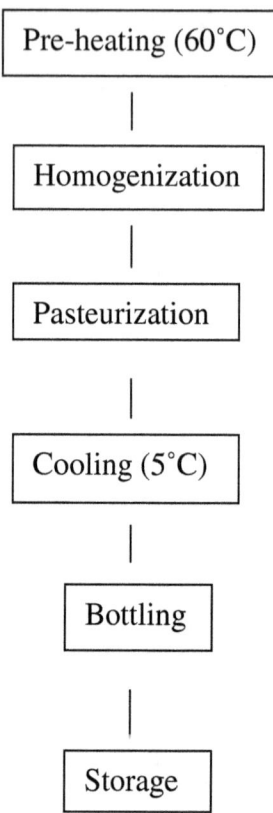

c) Flavored milk: Milk to which some flavors have been added to enhance consumer acceptability is called flavored milk.

Flow diagram for the manufacture of flavored milk:

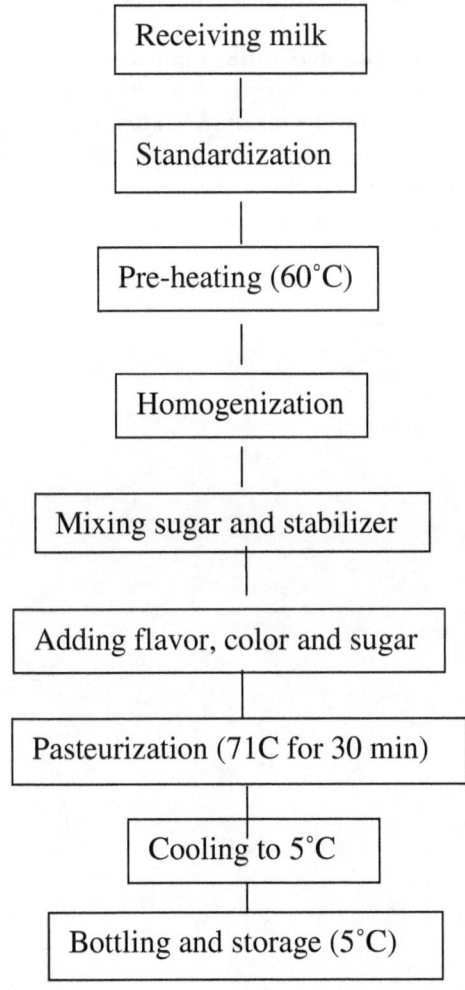

d) Vitaminized/ Irradiated milk: Vitaminized milk is milk in which one or more vitamins are added. Irradiated milk is milk in which vitamin D content has been increased by exposure to UV rays. Addition of vitamins into milk is called fortification and such milk is known as fortified milk.

e) Fermented milk: Fermented milk is prepared by employing one or more micro-organisms which brings change in the texture and flavor of milk after a well stipulated time. Lactic acid fermentation or souring of milk is the most important type of fermentation being carried out in milk. The various fermented milks include; yoghurt, dahi, buttermilk, kumiss, kefir etc.

Flow diagram for the manufacture of Dahi:

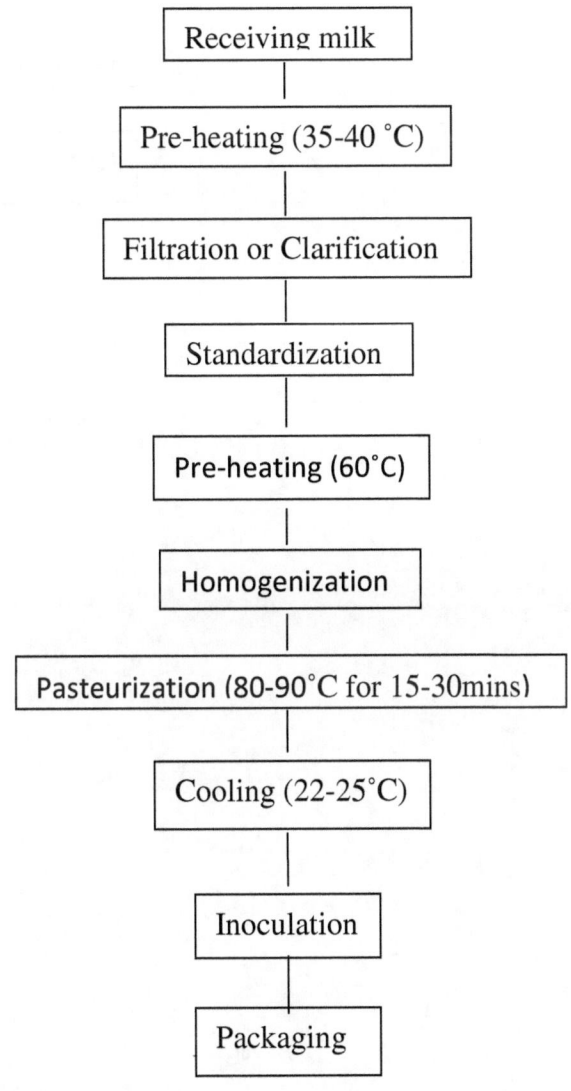

Receiving milk

Pre-heating (35-40 °C)

Filtration or Clarification

Standardization

Pre-heating (60°C)

Homogenization

Pasteurization (80-90°C for 15-30mins)

Cooling (22-25°C)

Inoculation

Packaging

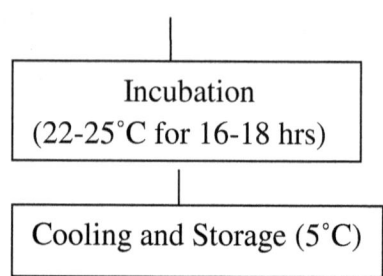

Incubation
(22-25°C for 16-18 hrs)

Cooling and Storage (5°C)

g) Standardized milk: Standardized milk is the milk whose fat content and/or solids-not-fat content have been adjusted to a certain pre-determined level. As per PFA rules, the Standardized milk should contain a minimum of 4.5 per cent fat and 8.5 per cent solids-not-fat.

Flow diagram for the manufacture of standardized milk:

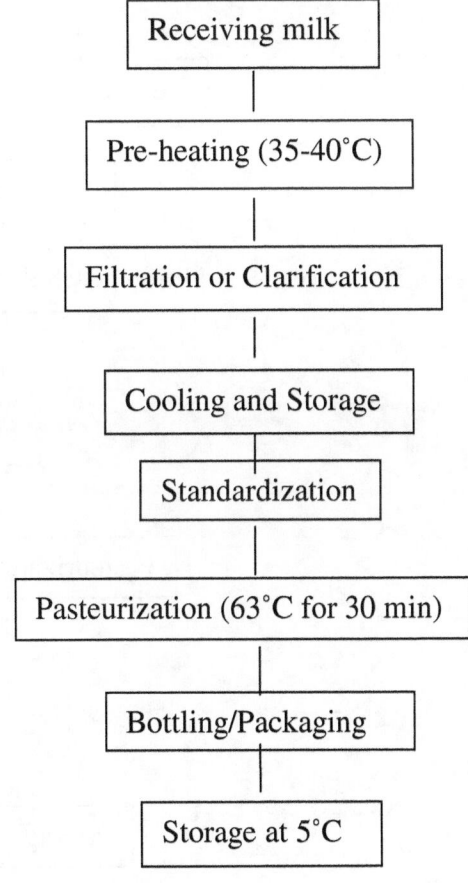

Receiving milk

Pre-heating (35-40°C)

Filtration or Clarification

Cooling and Storage

Standardization

Pasteurization (63°C for 30 min)

Bottling/Packaging

Storage at 5°C

h) Reconstituted/Rehydrated milk: Milk prepared by dispersing whole milk powder in water approximately in the proportion of 1 part powder to 7-8 parts water. It helps in meeting up the demands while facing shortage of fresh milk supplies in developing countries.

Flow diagram of manufacture of Reconstituted/Rehydrated milk:

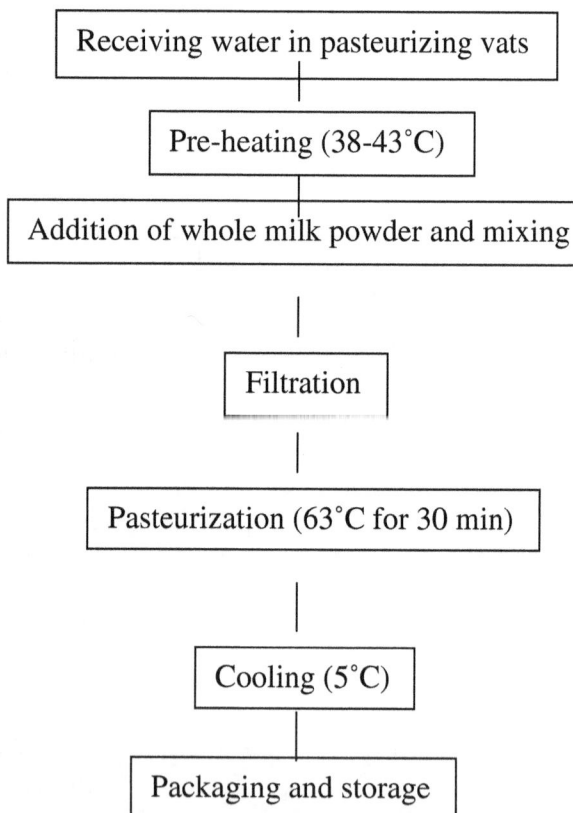

i) Recombined milk: Product obtained when butter oil, skim milk powder and water are combined in the correct proportions to yield fluid milk. This is also used for making up the supplies of milk during shortage in developing countries.

Flow diagram for the manufacture of Recombined milk:

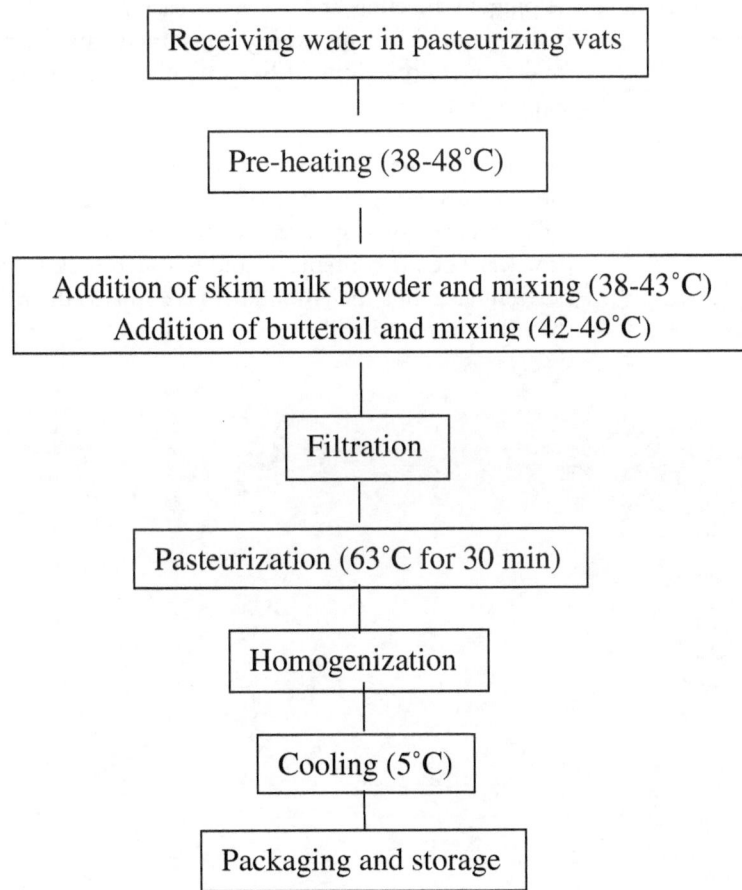

Receiving water in pasteurizing vats

Pre-heating (38-48°C)

Addition of skim milk powder and mixing (38-43°C)
Addition of butteroil and mixing (42-49°C)

Filtration

Pasteurization (63°C for 30 min)

Homogenization

Cooling (5°C)

Packaging and storage

j) Toned milk: The concept of preparing toned milk was developed in India. It is prepared by dispersing whole milk in water and skim milk powder and thus reducing its fat content to 3 per cent. It results in depreciation of milk prices by just reducing its fat content and at the same time increasing availability of milk for consumption. Toned milk is not only helping in meeting the ever increasing milk demands of the country but also being widely consumed with the perspective to intake low calorie, low fat but a highly nutritious diet in order to acquire fitness.

k) Double Toned milk: Double toned milk is based on the similar concept as that of toned milk. It is a special type of milk prepared by dispersing whole milk to water and skim milk powder. The fat content of double toned milk is adjusted to 1.5 per cent and thus the amount of fat and also the prices are lower than that of toned milk.

8. Milk Products

8.1. Cream: Cream refers to the fatty layer formed on the top of boiled milk when it is kept undisturbed at a low temperature.

According to PFA (1976), cream is the product of cow or buffalo milk or its combination which should contain not less than 25 per cent milk fat. Thus, cream may be defined as a part of milk which has been gathered to collect the major portion of milk fat. Table cream, coffee cream and light cream contains 20-25 per cent milk fat while whipping and heavy creams contain 30-40 per cent of milk fat. However, plastic cream contains comparatively higher percentage of milk fat with an average ranging from 65-85 per cent. Cream separation is a process being followed since time immemorial which involves two basic principles i.e., gravity separation method and centrifugal separation method. When milk is subjected to either gravity or centrifugal force, by virtue of the difference in their respective densities cream and skim milk gets separated from each other.

8.2. Butter: Butter is the fat concentrate obtained by churning cream and gathering it into a compact mass.

According to PFA Rules (1976), table butter is the product obtained from cow or buffalo milk or a combination thereof, with or without the addition of commonly used salt and annatto or carotene as coloring matter. It should be free from other animal fat, wax and mineral oils, vegetable oils and fats. No preservative except common salt and no coloring matter except annatto or carotene may be added. It must contain not less than 80 per cent by weight of milk fat, not more than 1.5 per cent curd and not more than 3.0 per cent common salt. Diacetyl may be added as a flavoring agent which if added should not exceed 4 ppm.

8.3. Butter oil: Butter oil is basically the fat-concentrate obtained mainly from butter or cream with the removal of all the water and solids-not-fat content. Raw material for the preparation of butter oil is cream. Butter oil is a rich source of milk fat among western dairy products. It is expected to be a rich source of fat soluble vitamin A and D. Butter oil contains nearly 99.5 to 99.8 per cent fat and only 0.1 to 0.3 per cent moisture.

8.4. Ice-cream: Ice cream is a frozen dairy product made by an appropriate blending and processing of milk, cream along with sugar and flavor with or without stabilizer or colour and with the incorporation of air during freezing.

According to PFA Rules (1976), ice cream is the frozen product obtained from cow or buffalo milk or either its combination or from cream, with or without the addition of cane sugar, eggs, fruits, preserved fruits, fruit juices, chocolate, nuts, dry fruits, edible flavors and permitted food colors. It may also contain permitted stabilizers and emulsifiers not exceeding 0.5 per cent by weight. The mixture should be suitably heated before freezing. The product should contain not less than 10 per cent milk fat, 3.5 per cent protein content and 36.0 per cent of total solids. However, when any of the aforesaid preparations contain fruits or nuts, the content of milk fat has to be proportionately reduced but should not be less than 8 per cent by weight. Starch may be added to a maximum extent of 5 per cent, with a proper declaration on its label.

Overrun in Ice-cream- After the processing of ice cream is done; the final weight shows an overrun which exceeds the initial weight of the mix. This overrun is usually defined as the volume of ice cream obtained in excess of the volume of the mix. It is expressed in percentage. This increased volume is composed mainly of the air incorporated during the process of freezing.

8.5. Cheese: Cheese is the milk product made up of curd obtained from milk by coagulating the casein with the help of rennet or similar enzymes in the presence of lactic acid produced by added microorganisms. The flavor and texture of cheese is attributed to curing or ripening i.e., holding it for a specific period of time under specified temperature and humidity conditions.

The cheeses are generally named after their place of origin. The most common, Cheddar cheese has been originated in Cheddar in England. Camembert and Roquefort cheeses have got their origin in France. Similarly, Swiss cheese in Switzerland and so on. On the basis of moisture content, cheeses are classified into very hard cheese (maximum 34% moisture), hard cheese (maximum 39-50% moisture) and soft cheese (50-80%). Cheeses may be ripened by bacteria (like *Lactobacilli, Pediococci, Brevibacterium* etc.) or by molds (*Penicillium spp.*) or may be unripened. The types of cheese and their ripening culture are given below:

Types of cheese and ripening culture	
Cheeses	**Ripening culture**
Cheddar	*Lactococcus lactis, Streptococcus cremoris, Streptococcus diacetyl lactis*
Swiss	*Propionibacteria shermanii*
Camembert	*Penicillium camemberti*
Cottage	*Lactococcus lactis, Leuconostoc species, Streptococcus cremoris*
Roquefort	*Streptococcus lactis, Penicillium camemberti*

Source: *Foods Facts and Principles* by Manay and Shadaksharaswamy (2008).

The composition of cheese varies as per its method of manufacture. Cheese has a high nutritive value. About 150g of cheese is equivalent in food value to one liter of milk. The average composition of some of the important varieties of cheese is given below:

Percent composition of different varieties of cheese (%)				
Name	**Moisture**	**Fat**	**Protein**	**Ash and Salt**
Camembert	47.9	26.3	22.2	4.1
Cheddar	36.8	33.8	23.7	5.6
Cottage	69.8	1.0	23.3	1.9
Roquefort	38.7	32.2	21.4	6.1
Swiss	33.0	30.5	30.4	4.2

Source: *Outlines of Dairy Technology* Sukumar De (2013).

8.6. Condensed milk: Condensed milks are the products obtained by evaporating part of water out of whole milk or skim milk with or without the addition of sugar.

According to the PFA Rules (1976), the various condensed milks have been specified as follows:

Unsweetened condensed milk (evaporated milk) is the product obtained from cow or buffalo milk or their combination, or from standardized milk, by the partial removal of water. It should contain not less than 8.0 per cent milk fat and not less than 26.0 per cent milk solids.

Sweetened condensed milk is the product obtained from cow or buffalo milk or their combination, by the partial removal of water and after addition of cane sugar. It should contain not less than 9.0 per cent milk fat, not less than 31.0 per cent milk solids and not less than 40.0 per cent cane sugar.

Unsweetened condensed skim milk (evaporated skimmed milk) is the product obtained from cow or buffalo milk or their combination by the partial removal of water. It should contain not less than 20.0 per cent milk solids and fat content should not exceed 0.5 per cent by weight.

Sweetened condensed skim milk is the product obtained from cow or buffalo skimmed milk or their combination by the partial removal of water and after addition of cane sugar. It should contain not less than 26.0 per cent milk solids and not less than 40.0 per cent cane sugar. The fat content should not exceed 0.5 per cent by weight.

8.7. Dried milk: Dried milk or milk powder is the product obtained by the removal of water from milk by heat or any other acceptable means, to produce a solid containing 5 per cent or less moisture. The dried product obtained from whole milk is called Dried Whole Milk or Whole Milk Powder (WMP) and that from skim milk is called as Dried Skim Milk or Skim Milk Powder (SMP). The various dried milk products and their Indian Standards are given below:

8.7.1. Milk drying systems: The system and process of milk drying is done by roller drying and spray drying method. In roller drying method, milk is applied as a thin film upon the smooth surface of a continuously rotating steam-heated metal drum. The film of dried milk is continuously scraped off by a stationary knife/ doctor blade located opposite the point of application of milk. The milk film has to be ground to obtain powder.

On the other hand, the basic principle of spray drying method consists of atomizing of milk to form a spray of very minute droplets which are directed into a large, suitably designed drying chamber, where they mix intimately with a current of hot air. Owing to their large surface area, the milk particles surrender their moisture instantaneously and dry into fine powder which is removed continuously.

As per the Indian Standards Institution (ISI), the specifications for dried milks are given in the table below:

Characteristics	Requirement for	
	Whole milk powder	Skim milk powder
Flavor and odor	Good	Good
Moisture (%) (Max.)	4.0	5.0
Total milk solids (%)	96.0	95.0
Solubility: Solubility index (Max) ml. Solubility (%) (Min) ml.	15.0(If roller-dried) 2.0 (If spray-dried) 85.0 (If roller-dried) 98.5 (If spray-dried)	15.0 (If roller-dried) 2.0 (If spray dried) 85.0 (If roller-dried) 98.5 (If spray-dried)
Total ash (%)	7.3	7.3
Fat (%)	Not less than 26.0	Not more than 1.5
Titrable acidity (%)	1.2	1.5
Bacterial count per g. (Max.)	50,000	50,000
Coliform count per g. (Max.)	90	90

Source: *Outlines of Dairy Technology* Sukumar De (2013).

9. Indian Dairy Products

The term 'Indian Dairy Products' refers to those milk products which solely originated in India. The importance of milk and milk products in India has been recognized since Vedic times in which milk is used for preparation of a variety of indigenous milk products. India is the world's largest producer of dairy products by volume and has the world's largest dairy herd. India accounts for more than 13% of world's total milk production and is also the largest consumer of dairy products in the world.

Prior to year 2000, India was not much noticed by International dairy companies as it did not participate actively in import or export of milk products. From the year 2000 Indian dairy products started having more presence on the global market. Milk production in India has developed significantly in the past few decades from a low volume of 17 million tons in 1951 to 110 million tons in 2009 and so on. Currently Indian dairy market is growing at an annual rate of 7%. Despite of the increase in production, a demand supply gap is on the way of our dairy industries due to a swift change in consumption patterns, fleeting urbanization and dynamic demographic patterns. Although India is moving ahead with a fine pace but still needs even faster to acquire higher standards for increasing the growth rate of dairy sector to match the pace of rapidly growing Indian economy. The various Indian or indigenous dairy products have been discussed in section 8.1.3.

10. The World Dairy Market and India

The world dairy market refers to the major importing and exporting countries of milk products and their trade practices in recent years. As per FAO statistics, world's aggregate milk production in the year 2002 was around 555 million tons. Milk production as compared to production in many other commodities is better distributed across the countries. There are only two countries, namely, India (13 per cent) and the United States (12 per cent) which account for more than 10 per cent of the world milk production; while countries in the European Union, Russian federation and Oceania account for around 22, 6 and 4 per cent of world milk production, respectively. World milk production in contrast to the domestic structure of milk production is dominated by cow milk. Cows and buffaloes account for around 85 and 10 per cent of world milk production; while goat, sheep and camels together account for less than 5 per cent of world milk production.

World trade in milk is limited; only 6 per cent of world milk production is undergoing trade worldwide. The perishable and bulky nature of milk is often cited as the prime reason for a limited trade of milk. However, the processed milk products such as milk powder, butter, and cheese have overcome the limitations. A comparison of world production and trade statistics for milk products indicate that out of the total world production of around 45 per cent of whole milk powder, 30 per cent of skim milk powder (SMP), 11 per cent of butter and 9 per cent of cheese are traded (FAO 1999). The world trade in milk products is therefore discussed with the trade figures for these products.

The Indian dairy products and their western counterparts are given in the table below:

Indian dairy product	Corresponding Western product	Principle of manufacture of Indian product
Concentrated whole milk products		
Kheer/ Basundi	Condensed milk	Partial dehydration in open pan with sugar, and occasionally rice, etc.
Khoa/ Mawa	Evaporated milk	Open-pan dehydration to a semi-solid consistency.
Rabri	Clotted cream	Partial dehydration in an open pan with sugar.
Kulfi	Ice cream	Concentrated milk, sugared and frozen.
(b) Coagulated milk products		
Dahi	Curd/ Yoghurt	Fermentation
Srikhand	Curd (sweetened)	Fermentation and straining, kneading with sugar.
Paneer	Soft cheese	Rennet coagulation, draining and salting.
Channa	Lactic coagulated green cheese	Acid coagulation and draining.
Products of the clarified butter fat industry		
Makkhan	Butter	Churning of fermented whole milk
Ghee	Butter oil	Clarification of butter or cream
Lassi	Buttermilk	By-product of makkhan
Ghee-residue	-	Heat-denatured SNF of butter or cream.

Source: *Outlines of Dairy Technology* Sukumar De (2013).

11. Indigenous Milk Products

a. Kheer- Kheer, also known as Basundi, is an Indian dessert prepared by the partial dehydration of whole milk in a karahi (a shallow, open, round bottomed pan, fitted with two loop handles) over a direct fire together with sugar and rice or occasionally semolina.

The average chemical composition (in percentage) of laboratory made Kheer under standardized conditions in a stainless steel kettle is as follows:

Moisture	Fat	Protein	Lactose	Ash	Sugar(added)
67.02	7.83	6.34	8.45	1.41	8.95

Source: *Outlines of Dairy Technology* Sukumar De (2013).

b. Khoa/Mawa- Khoa or mawa refers to the partially dehydrated whole milk product prepared by the continuous heating of milk in a karahi over a direct fire, while also constantly stirring–cum-scraping by using a ladle till it reaches a semi-solid consistency. According to PFA Rules (1976), khoa is the product obtained from cow or buffalo milk, or a combination undergoing rapid drying. The milk fat content should not be less than20 per cent of the finished product.

There are three main types of khoa viz., Pindi, Dhap and Danedar. These varieties differ in quality as well as in price. The classification of khoa is as under:

Type	Fat (%)	Moisture (%)	Specific sweets prepared
Pindi	21-26	31-33	Burfi, Peda etc.
Dhar	20-23	37-44	Gulabjamun, Pantooa etc.
Danedar	20-25	35-40	Kalakand, Gourd Barfi etc.

Source: *Outlines of Dairy Technology* Sukumar De (2013).

c. Rabri- This is an especially prepared concentrated and sweetened whole milk product, containing a number of layers of clotted cream. While the milk is slowly evaporated (without being stirred) at simmering temperature in a karahi over an open fire, the thin cream layers formed on the surface of milk are continuously broken up and moved to the cooler part of the karahi. When the volume of milk has been considerably reduced, sugar is added to it; then layers of clotted cream are immersed in the mixture and the finished product obtained by heating the whole mass for another short period. Rabri is quite popular in the northern and eastern regions of the country.

The composition of rabri depends on the initial composition of milk, concentration of milk solids and the percentage of sugar added. An approximate composition of rabri in percentage is given below:

Moisture	Fat	Protein	Lactose	Ash	Sugar(added)
30	20	10	17	3	20

Source: *Outlines of Dairy Technology* Sukumar De (2013).

c. Kulfi- Kulfi is an indigenous ice-cream frozen in small containers. For its preparation, milk is sweetened while being boiled with the addition of sugar and the product is concentrated to approximately 2:1. When the concentrate has been cooled malai (indigenous cream, formed on the surface of milk after boiling, cooling and left undisturbed), crushed nuts and flavor (commonly rose or vanilla) are added. The mix is then placed in conical or cylindrical moulds of different capacities made of galvanized iron sheets. The moulds are closed on the top by placing a small disc over them and edges made air tight by applying wheat-dough. Modern moulds are made up of plastics generally conical in shape with a screw cap on the top. The mix-in-moulds is frozen in a large earthen vessel containing a mixture of ice and salt in the ratio of 1:1.

d. Dahi- Indian curd or dahi is a well known fermented milk product consumed in a large scale throughout the country. According to PFA Rules (1976), dahi or curd is the product obtained from pasteurized or boiled milk by souring it by a harmless lactic acid or other bacterial culture.

It has been found that fermented milk products including dahi have more nutritious as compared to milk. Benefits of adding dahi to diet are cited below:

40

i. Dahi is more palatable and those who usually do not like drinking milk could consume it readily.

ii. Dahi could be more easily digested and assimilated by milk.

iii. Dahi exerts a possible therapeutic value and could be consumed during intestinal disorders to get rid of it.

The chemical composition of dahi in percentage is given in the table below:

Water	Protein	Fat	Lactose	Ash	Lactic Acid
85-88	3.2-3.4	5-8	4.6-5.2	0.70-0.72	0.5-0.11

Source: *Outlines of Dairy Technology* Sukumar De (2013).

e. Srikhand- It is a semi-solid, sweetish-sour, whole milk product prepared from lactic fermented curd. The curd (dahi) is partially strained through a cloth to remove the whey and thus produce a solid mass called chakka (a basic ingredient for srikhand). This chakka is mixed with the required amount of sugar to yield srikhand. The srikhand is further dessicated over an open pan to make the srikhand wadi sweet. The chemical composition of srikhand wadi in percentage is given in the table below:

Moisture	Protein	Fat	Lactose	Ash	Sugar	Lactic Acid
6.5	7.7	7.4	15.9	0.8	62.9	1.0

Source : *Outlines of Dairy Technology* Sukumar De (2013).

f. Channa- Channa, also known as paneer in certain regions of the country often used along with khoa for the preparation of indigenous sweetmeats (called Indian Mithai). According to the PFA Rules (1976), channa or paneer is the product obtained from cow or buffalo milk or a combination thereof by precipitation with acid like, lactic acid or citric acid. It should contain more than 70 per cent moisture and the milk fat content should not be less than 50.0 per cent of the dry matter. The average chemical composition of channa in percentage is given below:

Type of milk	Moisture	Fat	Protein	Lactose	Ash
Cow	53.4	24.8	17.4	2.1	2.1
Buffalo	51.6	29.6	14.4	2..3	2.0

Source: *Outlines of Dairy Technology* Sukumar De (2013).

g. Makkhan- Makkhan refers to the desi butter normally obtained by churning whole milk curd (dahi) with crude indigenous devices. According to the PFA Rules (1976), desi butter refers to the product obtained from cow or buffalo milk or a combination thereof without the addition of any preservative, including common salt, any other coloring matter or any added flavoring agent. It should be free from other animal fats, wax and vegetable oils. It should contain not less than 76.0 per cent of milk fat by weight. The chemical composition of makkhan is variable and depends on the method of manufacture. However, a standard quality makkhan (in percentage) meant for sale may confirm to the specifications given in the table below:

Moisture	Butter fat	Non-fatty solids	Lactic acid
18-20	78-81	1.0-1.5	Not more than 0.2

SOURCE: *Outlines of Dairy Technology* Sukumar De (2013).

h. Ghee- Ghee is the richest source of milk fat of all Indian dairy products. It may be defined as clarified butter fat prepared chiefly from cow or buffalo milk. According to the PFA Rules (1976), ghee is the pure clarified fat derived solely from milk or from desi cooking butter or from cream to which no coloring matter is added. Flow diagram for the manufacture of ghee through various methods, starting with milk is given below-

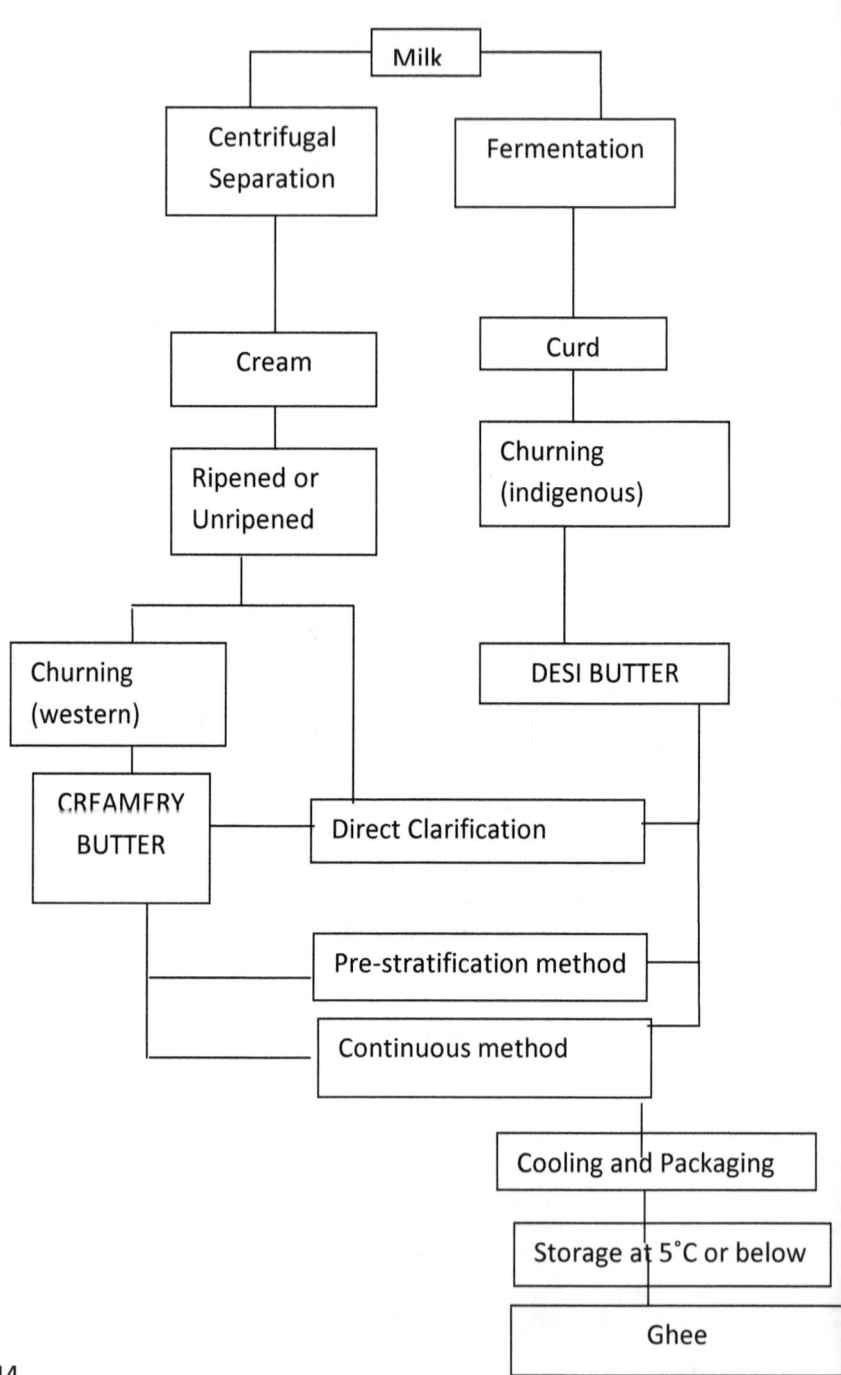

The chemical composition of ghee obtained from cow and buffalo respectively is given in the table below:

Characteristics	Requirements	
	Cow	Buffalo
Milk fat	99 to 99.5 per cent	
Moisture	Not more than 0.5 per cent	
Carotene (μg/g)	3.2-7.4	-
Vitamin A (I.U./g)	19-34	17-38
Tocopherol (μg/g)	26-48	18-37
Free fatty acid (% oleic)	Max. 2.8	
Charred casein, salts of copper, iron etc.	Traces	

Source: *Outlines of Dairy Technology* Sukumar De (2013).

    i. Ghee residue- Ghee residue refers to the charred (burnt) light to dark brown residue which is obtained on the strainer after the ghee is filtered. It is a by-product of the ghee industry. Ghee residue is a rich source of milk fat, proteins and minerals. Essentially it contains heat-denatured milk-proteins, caramelized lactose and varying amounts of entrapped fat, besides some minerals and water. Its color normally varies from light-brown to deep-chocolate brown. The color of the residue is chiefly influenced by the temperature-time combination. The higher the intensity of heat treatment, darker will be the color of residue and vice-versa.

j. Lassi- Lassi, also called chhas or matha, refers to buttermilk, which is the by-product obtained by churning of curdled whole milk for the production of makkhan or butter. Lassi contains appreciable amounts of milk proteins and phospholipids and is an excellent beverage for quenching thirst. It is widely served as a beverage in summers usually with the addition of ice, sugar or salt and often with or without additional flavours. The composition of lassi in percentage is given in the table below:

Water	Total solids	Fat	Solids-not-fat	Protein	Lactose	Ash	Lactic Acid
96.2	3.8	0.8	3.0	1.3	1.2	0.4	0.44

Source: *Outlines of Dairy Technology* Sukumar De (2013).

46

12. Milk Substitutes

Milk and other dairy products are expensive commodities and are not available as per the requirements in India and other developing countries. To overcome such situation, substitutes of milk and milk products are widely made from vegetable sources. These substitutes can be produced at a lower cost than natural milk by carefully formulating emulsifiers, minerals and vitamins in proper amounts. Fluid milk is being produced from soybean and groundnut. Excellent quality fluid milk can also be prepared from skim milk and vegetable fat. Such milk could also be pasteurized, homogenized and marketed just like ordinary milk. Substitutes of butter, ice-cream, coffee whitener etc can be obtained by proper processing and modification of vegetable fats.

Milk substitute for infants is also available in the market as Infant Milk Food which means the product prepared by spray drying of the milk of cow or buffalo or a mixture thereof. The milk may be modified by the partial removal/substitution of different milk solids; carbohydrates, such as sucrose, dextrose, maltose and lactose; salts like phosphates and citrates; vitamins A, D, E, B Group, Vitamin C and other vitamins; and minerals like iron, zinc, copper and iodine. Such method of adding minerals and vitamins in a product from an external source in order to enhance its nutritional value is called Fortification.

13. Packaging of Milk and Milk Products

Packaging means placing a commodity into a protective wrapper or container to facilitate transportation and storage. Packaging is an essential technique in which the most appropriate containers are used for identifying, protecting, carrying, and marketing of any product. It is a vital link between manufacturers and consumers for a safe delivery of products through various stages of manufacture, transportation, storage, and marketing. The main aim of packaging is to keep the food in a sound condition until it is in the hands of eventual consumers. Packaging is also necessary for attracting and encouraging the customers towards the product so that they tend to purchase it. Proper packaging is thus essential to achieve all these objectives. The importance of packaging could be summarized as follows:

Shelf life of the surplus food articles can be extended through adequate packaging and this allows the food to be distributed to different locations. By this, consumers may get more choices in terms of having variety in food availability. Moreover, rural producers may also be able to generate income from their surplus produce.

Proper packaging prevents the spoilage and wastage of food which may occur during transportation and storage.

Good packaging, attractive presentation and labelling encourage consumers to buy products.

Solutions to various problems regarding packaging differ from region to region. Such variations may result due to factors like availability of packaging materials, infrastructure, economy, climatic conditions and consumer habits. Foods with a shorter expected shelf-life like milk and milk products have different needs and may require more sophisticated packaging to protect it against air, light, moisture, bacteria and damage or spoilage during transportation.

13.1. Functions of packaging

Packaging plays a vital role in deciding the fate of the product. It should provide appropriate environmental conditions right from the time when the food is packed through to its consumption. Therefore, a good package should perform the following functions:

It should provide a perfect barrier against dirt and other contaminants so as to keep the product clean.

It should be able to prevent losses. For example, packages should be closed securely to prevent leakage.

It should protect food against physical and chemical damage. For example; protection against air, light, insects, and rodents.

The package design should be convenient for handling and transportation during marketing.

It should help the customers to identify the food article and also instruct them with its proper usage, manufacture date, expiry date, nutritional chart and other important informations.

It should persuade consumers to purchase the food item.

13.2. Development of Packaging Materials

Milk is a highly perishable liquid commodity and therefore requires proper keeping containers and packaging material at every stage of movement from the cow to the consumer so as to facilitate easy transport and avoid spoilage. In developed countries, all types of packaging machinery are available including those for the production of basic packaging materials and for converting these into finished packages, for filling and sealing, handling and storage, printing and testing etc. In many European countries town cow-keepers could still be found after the First World War but, for some reasons of hygiene and economy, they quickly disappeared. On the other hand, in developing countries this trend seems to be unavoidable for the dairy industry and will certainly be applied widely where town cow-keeping still exists.

The first significant development in the packaging of milk for retail sale came at the very end of last century with the introduction of the process for sterilized milk in which the retail container i.e., the glass bottles formed an integral and essential part. In 1930's bottling of pasteurized milk developed, firstly in America and soon after in Europe. The usage of glass bottle for milk remained unchallenged until 1933 when the first carton made of waxed paper was introduced. Later, the introduction of plastic material for packaging of dairy products in combination with paper, introduced the usage of cartons which were then used suitably for liquid milk. Presently a wide variety of packaging materials have came into existence which include paper-based products, glass, plastics, cartons, tetra packs, tin-plate, aluminium foil and laminates.

14. Production Policy and Regulation of Dairy Products

In India, earlier dairy production was basically a low input and low output system which used to affect the yield directly. As the consumer demand and market grew with time, dairy product prices and farmer's income continued to increase. Now the farmer's are gradually growing their herd sizes and aiming towards large scale production. Moreover, the private sector investors are partnering with dairy processors, funding for the construction of larger dairies and implementing various incentive schemes with a dual objective of increasing country's dairy output and also earning huge profits. Various incentive schemes include; the Ministry of Agriculture Research Program, Imports of Bovine Semen and Embryo, The National project for Cattle and Buffalo Breeding and other such schemes which focus on improving Indian cattle breeds. Private sectors and government institutions are also supporting farmers by providing training for veterinary care, artificial insemination services and other livestock management skills. Ministry of Health and Family Welfare has implemented Food Safety and Standards Authority of India (FSSAI) rules in the country in the year 2011 which consolidates various policies that aims at regulating food safety in India and sets the sanitary requirements for food safety, quality control, certification, packaging and labelling standards for all food products including milk and milk products.

Soon after India got independence in 1947, Milk Control Board was established to control the dairy supply and distribution chains. However, a number of problems arrived like middleman got hold of the sales profit as a result the actual producers started facing losses. In addition, since the processing units were established in cities hence the procurement and transportation of milk to rural areas got difficult. Consequently yield of milk declined and imports of milk powder went up. While the government was trying to deal with these problems a cooperative was set up in Kaira village at Gujrat with the objective of collecting, processing and marketing milk from it. Subsequently the Kaira Cooperative Union established a marketing agency named Gujrat Cooperative Milk Marketing Federation (GCMMF), with a three-layer structure of collection, processing and marketing of dairy products at village, district and state level. The district units also supported the milk producers in this by providing technical support such as providing cattle feed, veterinary care, artificial insemination, education and training. Today the milk cooperatives of Gujrat own the GCMMF which is one of the largest food product business in India. GCMMF is also the largest exporter of dairy products from India which owns the brand AMUL. Later in 1965, National Dairy Development Board (NDDB) was set up which proved to be a blueprint for a dairy revolution across the country popularly known as 'Operation Flood'.

15. Operation Flood Era

The dairy sector of India witnessed a spectacular growth between 1971 and 1996. This period is known as Operation Flood Era. The main aim of establishing this integrated cooperative program was to develop Indian dairy industry. It basically worked in a three phase system with NDDB designated by the government of India as an implementing agency. The major purpose of its establishment was to assure year round market to the rural milk producers and also to establish a strong connecting link between the rural producers and urban markets by using modern technologies. The Operation Flood is one of the biggest rural development programs which helped India to come forth and prove itself to the world in terms of its milk production skills. In addition to this, with this program nearly 10 million farmers enrolled themselves as members in around 73000 milk cooperative units. This program ran for about 27 years and towards its end milk production in India stretched its feet from 21 million tons in 1970 to about 69 million tons in 1996.

16. Development and Future Outlook

Dairy industry is crucial for the overall development of India. Today India not only enjoys a higher rank among the list of high milk producing countries but is also a world's largest milk consuming nation. On an average, an Indian family allocates 15-18 per cent of its monthly food expenditure on milk and milk products. Dairy products are comparatively cheaper source of nutrition for millions of families in India. Not only this but also milk and milk products are fairly acceptable source of protein to a large segment of vegetarian population in India. As the income of common man is increasing continuously it has been predicted that milk demands are going to rise faster in the coming decades which may also exceed the overall milk production. Increasing income of people and the ever enhancing GDP rate have influenced milk demands both in rural as well as urban sectors. Apart from this, other reasons like costly cattle feed and decreased availability of labor in rural areas have resulted in increased cost of production. It has been estimated that in the near future demands for milk are going to increase at a rate of 7 per cent per annum if the current scenario persists.

Following steps could be taken to overcome problems which are being faced along the way of development in Indian dairy industry:

Reducing the cost of production – Efforts need to be made in order to increase milk production and subsequently reducing the cost of production. This could be achieved by introducing high yielding milch animals, providing animal breeding facilities and giving proper trainings regarding animal health care to the farmers and workers. For this, the dairy industries, veterinary science institutions and government will have to take some firm steps towards this direction.

Infrastructure development – Indian dairy industry needs to develop proper infrastructure in order to carry forward the dairy industry in terms of production, processing and marketing so as to meet international dairy standards. Strategies must be implemented for the production of safe and good quality products alongwith a suitable legal backup.

Focus on Specialty Dairy Products – Indian dairy industry is unique in terms of availability of specialty dairy cuisines and with the availability of not only cow milk but also buffalo milk. Both these milk varieties can be successfully processed for the preparation of a variety of indigenous dairy products which are not only liked but also cherished by the customers.

Bibliography

Chand, S. ; Saraiya, A. and Sridhar, V. (2010). Public Private Partnership in Indian Dairy Industry (pdf).

D. K. Thompkinson, D.K. Gahlot and O. N. Mathur (1976). *Indian Journal Dairy Science.* 29(4), 316 pg.

FAO Corporate Document Repository. *Agriculture and Consumer Protection.*

Karmakar & Banerjee (2006). Indian Dairy Industry. *IUF Dairy Industry Research.*

PFA (1976)and FSSAI(2011) Rules.

Shakuntala Manay and Shadaksharaswamy (2008). *Foods Facts and Principles.* (p 303-324). New Age International Publishers.

Sukumar De (2013). *Outlines of Dairy Technology.* (539 pages). Oxford Univ. Press.

A. Multiple Choice Questions

1. Casein Present in Milk is found in the form of
   a. Magnesium caseinate-phosphate complex
   b. Calcium caseinate-phosphate complex
   c. Potassium caseinate-phosphate complex
   d. None of the above

2. .... is the basis for checking pasteurization efficiency of milk
   a. Peroxidase and catalase test
   b. Phosphatase test
   c. Analase test
   d. None of the above

3. Clot-on-boiling (COB) test is carried out to
   a. Determine the bacterial count of milk
   b. Determine the heat stability and pH of milk
   c. Determine the heat stability of milk
   d. None of the above

4. Rind Rot is observed in
   a. Butter
   b. Milk
   c. Yogurt

d. Cheese

5. Ripening culture involved in the formation of camembert cheese is/are
   a. *Streptococcus thermophilus*
   b. *Lactobacillus bulgaricus*
   c. *Both a and b*
   d. *Penicillium camemberti*

6. Which of the following do affect the composition of milk
   a. Breed
   b. Interval of milking
   c. Stage of lactation
   d. All of the above

7. Phosphatase test is not applicable for
   a. Pasteurized milk
   b. Sterilized milk
   c. Both (a) and (b)
   d. None of the above

8. The only carbohydrate present in milk is
   a. Maltose
   b. Glucose
   c. Sucrose

d. Lactose

9. The most variable component of milk is

   a. The protein followed by fat

   b. The fat followed by protein

   c. The protein

   d. All of the above

10. The fat content of toned milk should be

    a. Not less than 1.5%

    b. Less than 3%

    c. More than 1.5% but less than 3%

    d. Not less than 3%

11. Cheese is classified on the basis of

    a. The fat content

    b. The moisture content

    c. The protein content

    d. Bothe a and b

12. National Dairy Development Board (NDDB) was established in

    a. 1966

    b. 1965

    c. 1976

    d. 1956

13. The head office of NDDB is located at

    a. Delhi

    b. Mumbai

    c. Anand

    d. Banglore

14. As per PFA rule the fat content of khoa should not be less than   [ICAR '05]

    a. 10%

    b. 20%

    c. 30%

    d. 40%

15. The energy value of cow milk is

    a. 25 C/100 g

    b. 35 C/100 g

    c. 75 C/100 g

    d. 90 C/100 g

16. Milk is low in

    a. Iodine

    b. Iron

    c. Copper

    d. All of the above

17. Which of the following aspect of milk decides its market price?

    a. Protein

    b. Fat content

    c. Solid content

    d. None of the above

18. Specific gravity of lactose is (60°F)

    a. 1.021

    b. 1.999

    c. 2.233

    d. 1.510

19. Specific gravity of of skim milk is

    a. More than cow milk

    b. Less than cow milk

    c. More than buffalo milk

    d. Less than water

20. Which of the following come under Platform test?

    a. Lactometer reading

    b. Acidity

    c. Appearance

    d. All of the above

21. As per legal standards, the fat content of standardized milk should be

    a. 2.5%   b. 3.5%

    c. 4.5%   d. 5.5%

22. Yoghurt is similar to

    a. Indian dahi

    b. American matzoon

    c. Egyptian leben

    d. None of the above

23. The fat percent of double toned milk is

    a. 0.5   b. 2.5

    c. 1.5   d. 4.5

24. The diacetyl content in butter should be

    a. 1 ppm   b. 2 ppm

    b. 3 ppm   d. 4 ppm

25. Which of the following cheese is prepared by using skim milk

    a. Cottage

b. Cheddar

c. Roquefort

d. Camembert

26. Yellow color of cow milk is due to the presence of

   a. Xanthophyll

   b. Carotene

   c. Riboflavin

   d. Bixin

27. pH of fresh cow milk is

   a. 6.0

   b. 6.2

   c. 6.4

   d. 6.6

28. Which of the following creams have 20-25% fat content

   a. Table cream

   b. Coffee cream

   c. Light cream

   d. All of the above

29. Ministry of Health and Family Welfare implemented FSSAI rules in the country on the year

a. 2007

b. 2009

c. 2011

d. 2013

30. Whipping cream contains .... % fat

a. 10%

b. 20%

c. 30%

d. 40%

B. Descriptive type questions

1. Define milk? Describe the composition of milk in detail.

2. What are platform tests? Why is it necessary to perform these tests while receiving milk in a dairy industry?

3. Give a detailed description of special milks. What PFA standards have been assigned for different classes of milk in India?

4. Write a short note on the role of dairy industry in development of our country. What vital steps should be undergone to improve the existing scenario of Indian dairy?

5. Why is packaging of milk and milk products essential? Discuss the functions of packaging in brief.

Answers to Multiple Choice Question's

1. (b)
2. (b)
3. (c)
4. (d)
5. (d)
6. (d)
7. (c)
8. (d)
9. (b)
10. (d)
11. (b)
12. (b)
13. (c)
14. (b)
15. (c)
16. (d)
17. (b)
18. (b)
19. (a)
20. (d)

21. (c)

22. (a)

23. (b)

24. (d)

25. (a)

26. (b)

27. (d)

28. (d)

29. (c)

30. (d)